ARE WE THE ALIENS?

A Look into Our Past (Human Life on Earth and Other Planets)

BY
PAUL SIMS

TABLE OF CONTENTS

Introduction

This book is focused on facts which reveal evidence of a technologically advanced and highly intelligent human civilization which existed in our pre-historic past. Many scientists believe that humans are not from earth. Did humans come to this solar system and terraform the earth, Mars, and possibly the planet beyond Mars which is now the asteroid belt? Was there an ancient war between the civilizations on Mars and the planet beyond?

The star Sirius and Our sun are a binary system. These two stars are at their closest every 26,000 years. This cycle follows what are said to be golden ages, a time where the human civilization is at a zenith. Is it possible that humans came here from Sirius and or other star systems?

The Earth

Our planet is known to have had three distinct atmospheres over the course of its life so far. Earth's first atmosphere is said to have consisted of hydrogen, helium, and other gasses. Solar wind is thought to have changed this atmosphere. Earth developed a magnetic field that protects the earth from this solar wind. The second atmosphere formed soon after, about 4.5 million years ago. Volcanic outgassing released water vapor, carbon dioxide, methane, ammonia, and other gasses. When the earth cooled these gasses and water vapor formed clouds and oceans. Carbon dioxide and water vapor also created oxygen by way of photolysis and ultraviolet radiation. The third atmosphere was formed approximately 2.5 million years ago. By this time more oxygen has formed due to oceanic organisms called cyanobacteria. The current atmosphere contains about 78% nitrogen, 21% oxygen, 1% argon and the rest mostly carbon dioxide and other gasses due to plant life.

Looking at a timeline of life on earth, 3.5 billion years ago the oldest prokaryotic fossils are found. 2.5 billion years ago oxygen starts to accumulate into the atmosphere. At 2.1 billion years in the past the oldest eukaryotic fossils. 700 million years ago basic multicellular organisms. 420 million ago plants start to spread throughout the world. 370 million years, amphibians appear. 360 million years back the first insects, 340 years reptiles, then at 280 years in the past there was a mass extinction of the life on earth at that time. The cause of the mass extinction is unknown. Closely following the mass extinction, the first dinosaurs are found to have come into being at 230 million years ago. 195 million yeas back birds arrive on earth, 141 million years the first flowering plants, then at 66.4 years ago there was another mass global extinction of life on the planet earth. This global extinction is thought to be caused by an asteroid that hit the earth at that time. 2.4 million years in the past there was an ice age. This leads us into our modern era. While the human race has only occupied 2 million years on the timeline, the dinosaurs lived for about 175 million years on earth before their extinction. The Cretaceous-Tertiary extinction event or the K-T event is believed to have cause the global extinction of the dinosaurs. The Chicxulub crater in the Yucatan Peninsula, Mexico is thought to be the asteroid which caused this global extinction period.

There are approximately 391,000 plants on the earth, 1-2 million different animal species, 900-thousand types of insects, 332.5 million cubic miles of water or 1,386 million cubic kilometers of water on earth which is equal to 326 million trillion gallons. How did all this water and such an overabundance of life come to exist on the planet earth? While every other planet that we know about is barren, the abundance of life on earth seems highly unlikely to have arisen on its own by chance.

How did earth form a perfect climate with a perfect amount of oxygen for humans? The earth's radius is 3,958.8 miles. One day on earth is 23.56 hours, one year is 365.2422 days. The distance to the sun is 92.96 million miles.

Earth's mass is approximately 5.972 × 10^24 kg. The planet is composed of 32.1% iron, 30.1% oxygen, 15.1% silicon, 13.9% magnesium, 2.9% sulfur, 1.8% nickel, 1.5% calcium, 1.4% aluminum, and 1.2% other elements. The core of the planet is thought to be about 88% iron and the rest nickel and sulfur with less than 1% trace elements.

The layers of earth starting with the outer layer is the crust, upper mantle, mantle, outer core, and inner core. The crust is an average of 40 km thick. Igneous, sedimentary, and metamorphic rocks are found in the crust. The mantle is 500 to 900 degrees Celsius. It is about 2,900 km thick. The rock is similar to the crust. The outer core is mostly iron and nickel in liquid form. It is approximately 5,000 to 7,000 degrees Celsius. It is about 2300 km thick. The inner core tempeture is about 9,000 to 13,000 degrees Celsius. The very core of the earth is said to be under so much pressure that it remains in solid from. It is about 1,200 km thick. Scientists also think that there may be magnetized crystals which are oriented to the north and south, surrounding the inner core.

The atmosphere of earth is approximately 78% nitrogen and 21% oxygen with the rest argon, carbon dioxide and other trace elements. The layer of atmosphere closest to the earth's surface is called the troposphere. It extends to about 10 to 15 km above the surface of the planet. Most of the planet's weather happens in this region. It is the largest layer of the atmosphere and accounts for 75% of the earth's overall atmosphere. The stratosphere is the second layer of the atmosphere, it is about 15 to 50 km above earth's surface. The upper layer of the stratosphere absorbs more of the sun's

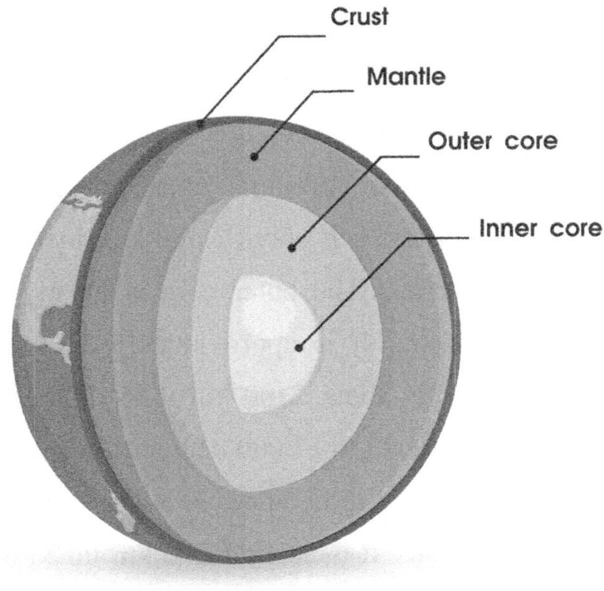
Crust
Mantle
Outer core
Inner core

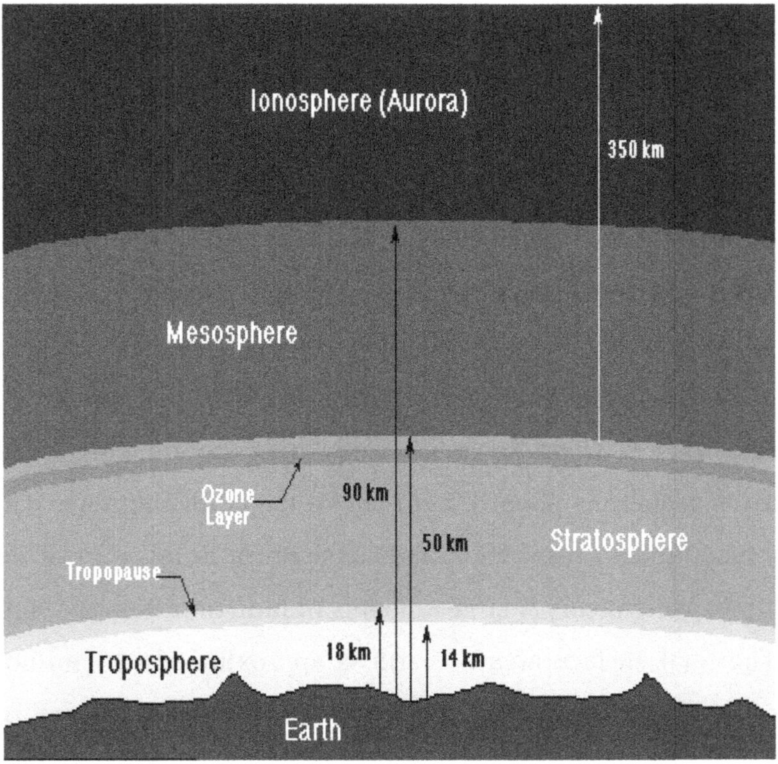
Ionosphere (Aurora)
350 km
Mesosphere
Ozone Layer
90 km
50 km
Stratosphere
Tropopause
18 km
14 km
Troposphere
Earth

radiation by the ozone layer, which leads to a higher tempeture in the upper region than the lower region of the stratosphere. The ozone layer is a part of the stratosphere which contains a protective layer of three oxygen molecules called ozone. This layer shields earth from harmful ultraviolet radiation from the sun. The mesosphere is about 50 to 80 km above the earth's surface. The tempeture is cooler in this part of the atmosphere. It is the part of the atmosphere where most asteroids burn up on entry. The thermosphere is about 80 km above the earth's surface. The tempeture can reach thousands of degrees as this part of the atmosphere receives large amounts of radiation from the sun. The total surface area of earth is approximately 15 million square km. About 71% of earth's surface area is covered by ocean water. The average atmospheric pressure at sea level is about 101.33 kilopascal per square inch. The average salinity of the earth's ocean is around 3.5% salt.

Earth is surrounded by a magnetic field that is generated with a dynamo process in the earth's core. At the surface of the earth this magnetic field creates dipoles at the north and south poles of the planet. This magnetic field extends out to create what scientists call the magnetosphere. The magnetosphere is full of charged particles which follow the magnetic field lines of earth. The Van Allen radiation belts are a part of the magnetosphere. This part of the atmosphere helps to protect earth from the sun's radiation. When some of the sun's radiation gets through the magnetosphere, atoms are excited and ionized creating the aurora borealis.

The orbital speed of earth is about 29.8 km per second. This is fast enough to cover the distance from the earth to the moon in about 3.5 hours. The earth's axis has a tilt of 23.4 degrees. The tilt causes seasonal climate changes. Sumer happens when the Tropic of Cancer faces the sun. Winter in the southern hemisphere happens when the Tropic of Capricorn faces the sun. During the Summer solstice the sun rises north of true east and sets north of true west. In the Winter this is reversed. In the northern hemisphere Winter solstice is on Dec 21st and Summer solstice is on is on June 21st. Spring equinox is around March 20th and autumn equinox around Sept 22. It is the opposite for the southern hemisphere. Distance from the earth to the sun, orbit, rate of rotation, axial tilt and magnetic field all contribute to earth's self-regulating environment and the stability of our habitat.

The earth's core, atmosphere, and eco systems are so complex in their make-up, mechanics, and function that it seems like they were designed. Earth's orbital speed, tilt, gravity, tempeture, atmospheric pressure, distance from the sun, it's abundance of life, perfect oxygen levels, and all the other things that lend such an appeasing habitat for human life, seems all too perfect. Could the earth and maybe other planets in our solar system have been terraformed for human life?

KT boundary

The Cretaceous boundary or the Cretaceous Tertiary is a thin layer of clay with a strong concentration of iridium that is found all over the earth. Radiometric dating puts it at 66 million years ago. It is known to mark the end of the Mesozoic Era and the beginning of the Cenozoic Era. The K-T boundary is thought to be evidence of the mass extinction of the dinosaurs and the Cretaceous Paleogene extinction event. The event is said to have made the world oxygen levels drop drastically. Scientists think that it was caused by a meteor hitting earth. The Chicxulub Crater in the Yucatan Peninsula which happened approximately 66 million years ago is thought to be evidence of this meteor hit. Some believe that the event may have been caused by multiple impacts and mention other impact sites that could be responsible for the event such as the Shiva Crater off the west coast of India or the Boltysh Crater in Ukraine and the Silverpit Crater in the North Sea.

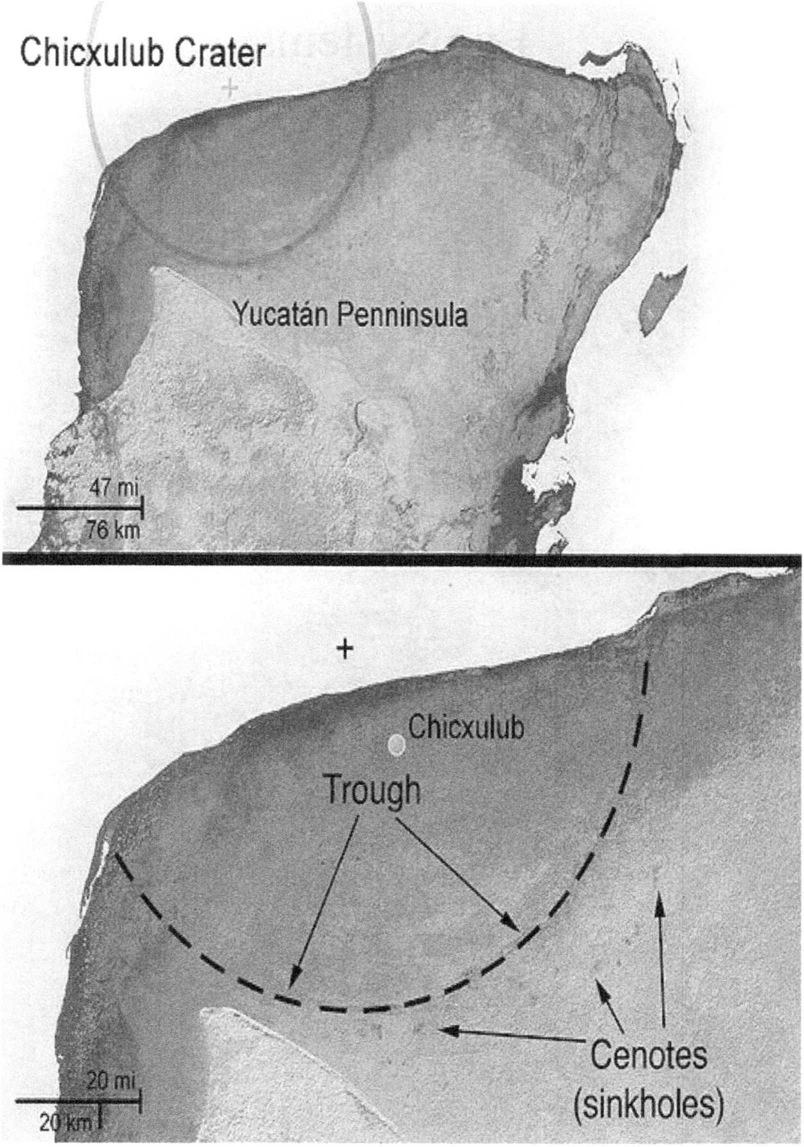
Chicxulub Crater
Yucatán Penninsula
47 mi
76 km
Chicxulub
Trough
Cenotes
(sinkholes)
20 mi
20 km

Eco Systems

An eco-system is a group of organisms and environments interacting with one another. Ecosystems are present on many levels, from the microscopic all the way up to the planetary scale, cycling nutrients to support life. On earth we have terrestrial and aquatic ecosystems. They include forests, tundra, deserts, grasslands, fresh water and marine. Then there are sub-sets of these as well such as: Tropical Rainforest Ecosystems, Temperate Forest Ecosystems, Boreal Forest Ecosystems, Taiga Ecosystems, Stillwater Ecosystems, River and Stream Ecosystems, Littoral Zones, Coral Reefs, Polar ecosystems, Mountain Zones, Estuary Ecosystems, Saltwater Wetland Estuary Ecosystems, Ocean Floor Ecosystems, Lakes, Bays, Gulfs, Oceanic ecosystem, and more.

Humans

DNA is a complex code of information. The more that is found out about the DNA, the more we see that it is of intelligent design. This kind of design is far too complex to have happened by chance. Logic brings us to two conclusions; DNA was created by someone. The other conclusion: we did not originate on this earth by means of natural selection or evolution. We came from somewhere else.

Let's back up a little and take a closer look at the origins of the human. In 1953 Sir Francis Crick and James Watson discovered DNA. Crick claims that DNA is too complex to have been developed randomly. He also believed that DNA was not from earth. DNA has an RNA genetic code that is very complex and very exact. It resembles computer coding. If any of this code is wrong, development will not take place. These strands of code create all parts of the human body, and the origin of these codes are completely unknown. DNA seems to have been designed and carefully engineered by a high intelligence.

As we now know evolution has many problems. It is important to note that evolution can only apply after creation has begun, so evolution does not explain creation itself. Evolution cannot explain how a living cell can originate from non-living matter (chemical attraction). In his book "Darwin's Black Box", Dr Michael Bebe focuses on what he calls an irreducible complexity. Bebe states that parts of the human body are irreducible. Meaning that removing just one part of the system, will render the whole system itself unfunctional. Bebe uses the bacterium flagellum as his example. This flagellum needs all its parts to function and could not have developed as evolution states that it should. A complex system like the flagellum had to have been whole and complete from the beginning according to the scheme of evolution. Evolution states that development can only happen out of necessity, and every part that is developed must serve a purpose for its development. For a system that is an irreducible complexity, this will not happen. It turns out that the human body itself is irreducibly complex. That is, if you take just one part out of the human body, it cannot function. This means that the human body must have been whole and complete from the start, or at least the program in the DNA had to have been complete from the beginning.

It also suggests that humans did not evolve on earth and may have been brought here by another species, or humans came here on their own behalf.

The Flagellum

A flagellum is an appendage that protrudes out of the cell body. Eukaryotic cells, for example have a flagellum that looks like a tail. This flagellum serves as a propulsion mechanism for eukaryotic cells like the sperm cell. The prokaryotic flagellum is essentially a rotary motor. The bacterial flagellum has a rotary motor made of protein. The rotor is a 20-nanometer thick tube with a shaft which passes through 4 rings and is attached to bearings and gears all made of proteins. The motor is driven by either a sodium ion or a proton pump. This all allows the cell to achieve roughly 60 cell lengths per second. A cheetah, in comparison, achieves about 25 body lengths per second. Some scientists say that the flagellum is an irreducible complexity. This means that evolution could not have created the flagellum. All parts of the flagellum are needed for it to function, and it would have been made whole and complete. It could not have developed over time. Other scientists argue that proteins can be taken away and the flagellum still functions. Many argue back saying that you can take away the brakes, steering, cooling system and more from a car but it will still run. Given that the car will still run, it will not run as intended and could never serve its original purpose in any effective or efficient manner.

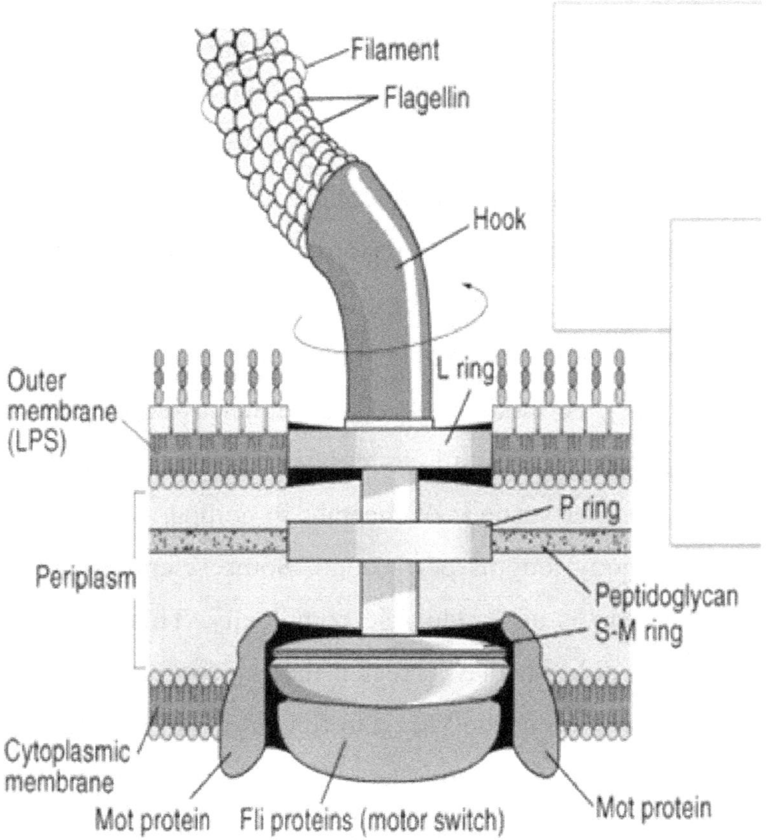
Filament
Flagellin
Hook
L ring
Outer membrane (LPS)
P ring
Periplasm
Peptidoglycan
S-M ring
Cytoplasmic membrane
Mot protein
Fli proteins (motor switch)
Mot protein

In 1779 Johann Friedrich Blumenbach classified humans into five distinct races: Ethiopian (Black), Caucasian (White), Mongolian (Asian), American (Indian), Malayan (Brown). Later in 1962 anthropologist Carleton S. Coon, also classified humanity into five races: Caucasoid (White), Congoid (Negro), Capoid (Bushmen), Mongoloid (Asian), and Australoid (Aborigine). Sub-dividing the White race started to happen adhering to differences in hair color, eye color and skin tone. Blonde hair, red hair, brown hair. Blue, grey, green and brown eyes. White skin and olive skin tones, with all these differences taken into consideration.

Today's Homo sapiens (modern humans) arose 300,000 to 200,000 years ago. Homo genius and Homo habilis lived just over 2 million years ago. Around this same time Homo erectus spread throughout Africa. Erectus is thought to be the first human to control fire and live as a hunter gatherer. Evidence reveals that Homo sapiens and Homo neanderthalensis (Neanderthal) emerged around the same time. In Africa and Europe Homo sapiens and Neanderthal interbred.

It is unclear as to which species gave rise to Homo lineage. There are four groups of Australopithecines: Garhi, Sediba, Africanus and Afarensis. These species and Neanderthal did not give rise to Homo genius. It has also been discovered that Homo habilis and Homo erectus coexisted which means that habilis did not give rise to erectus. Homo sapiens are known to have interbred with both Neanderthals and Denisovans in Africa and Eurasia. Scientists also found a sub-Saharan African race called Homo heidelbergensis with an unknown lineage. The three mainly accepted species in human taxonomy are: Homo habilis, Homo erectus and Homo sapiens.

Several different Homo species lived at the same time and even interbred. Also, there is no known race that gave rise to Homo sapiens. The origin for Homo sapiens (and other Homo species) is unknown. Did the Homo species come here from somewhere else?

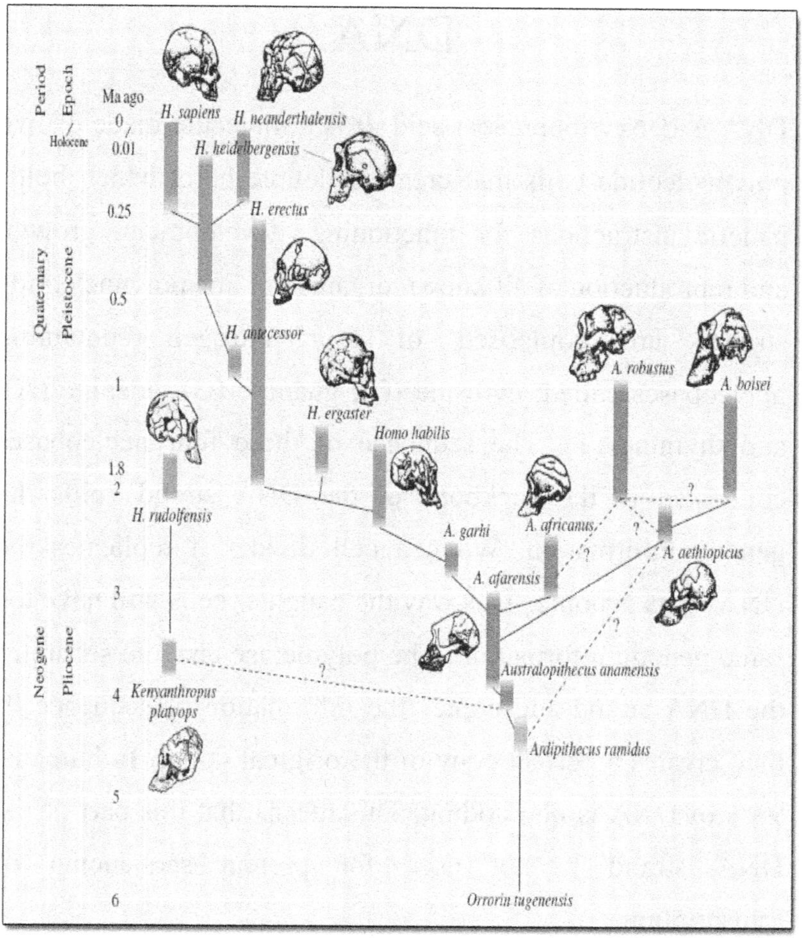
Period
Epoch
Ma ago
0
Holocene
0.01
0.25
0.5
1
1.8
2
3
4
5
6
Quaternary
Pleistocene
Neogene
Pliocene
H. sapiens
H. neanderthalensis
H. heidelbergensis
H. erectus
H. antecessor
H. ergaster
Homo habilis
H. rudolfensis
A. robustus
A. boisei
A. garhi
A. africanus
A. aethiopicus
A. afarensis
Australopithecus anamensis
Kenyanthropus platyops
Ardipithecus ramidus
Orrorin tugenensis

DNA

DNA is deoxyribonucleic acid. It is a molecule made of two polynucleotide coils that create a double helix which holds genetic instructions for functioning, development, growth, and reproduction of all known organisms. The polynucleotide strands are composed of four nitrogen containing nucleobases called: cytosine (C), guanine (G), adenine (A), and thymine (T). The sequence of these four nucleobases connected to the backbone of the DNA strand holds the genetic information. When a cell divides it replicates the DNA in its genome. This way the daughter cells will have the same genetic information. The polymerase enzyme separates the DNA strand and copies the information or sequence. It then creates a perfect copy of the original donor. In humans, 98% of DNA is non-coding. This means that this part of the DNA strand is not used for protein sequencing or transcription.

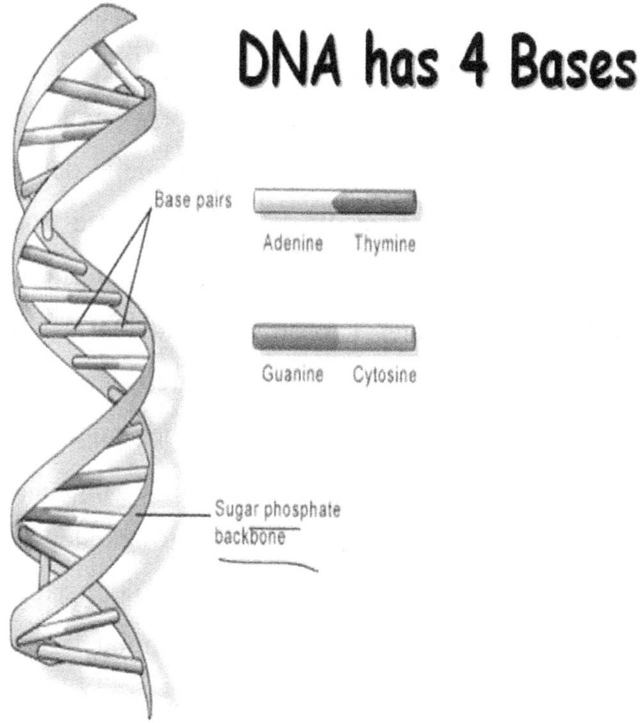
DNA has 4 Bases
Base pairs
Adenine
Thymine
Guanine
Cytosine
Sugar phosphate
backbone

Not only is the origin of DNA unknown, the functionality and complexity are a mystery. The man who discovered DNA, Sir Francis Crick, did not believe that DNA was from this earth. He believed that DNA was too complex to have evolved or developed on its own. Crick wrote in detail on the subject in his book 'Life Itself'. The genetic code is a complex product of intelligence, and evolution cannot explain the genetic code. The only conclusion is that DNA and genetic coding had to have been created. The fact that all living organisms contain this coding further suggests that all life is based on the same template. This means design. Another sequence that is imbedded into existence is the universal mathematical golden ratio. The golden ratio is also a universal template of creation. These templates of creation tell us that not only life but all of existence was engineered.

RH- Blood

In between Spain and France is the Pyrenees Mountain range. The people in that region are the Basques. Their language is not related to any other in that part of the world. Up to 35% of them have Rh- blood type. Human blood types include O, A, B, AB and Rh-. 85% of humans are Rh+ and can receive blood from positive or negative blood donors, but Rh- humans cannot. It is fatal for a person who is Rh- to receive donor blood that is Rh+. People who have Rh- blood type commonly have slower hart rate, lower blood pressure and in some cases, an extra vertebra in the spinal column. Rh- and Rh+ blood types cannot procreate without medical intervention. When women who have Rh- blood get pregnant with a man who is Rh+, the woman's immune system can kill the baby.

Rh- blood has no Rhesus factor antibodies on blood cells and no antibodies in the plasma. An antigen causes the body to produce an antibody. These antibodies are proteins. If there is an Rh+ father and a Rh- mother and the baby has Rh+ blood the situation can lead to hemolytic disease. The introduction of Rh+ blood cells will cause the mother's body to create Rh+ antibodies and the antibodies will attack the next baby. The mother can only prevent her body from devolving the Rh+ antibody by medical intervention. This suggests that there are two different types of modern-day humans alive, walking the earth side by side today. The Rh- human cannot procreate with the Rh+ human naturally without medical intervention. This means that these two types of humans did not have the same place of origin. They are two completely different types of humans.

The Rhesus blood group consists of 49 blood group antigens. Among these groups is the Rh(D) blood type. Rh(D)+ blood means the blood has the Rh(D) antigen on the red blood cells. Rh(D)- blood means that there are no Rh(D) antigens on the surface of the red blood cells. By population approximately 7% of African Americans are Rh(D) negative, 8.8% of Saudi Arabians are Rh(D) negative, 9.5% of Moroccans, 11% of Ethiopians, 15% in United States, 16% of Europeans, 29% of High Atlas Moroccans, and 30% of the Basques.

Where did humans come from if they didn't evolve here? Many cultures have stories of where they come from. Indians, Asians, Blacks, Whites. Did the races of the world all come from the same place, or did they come from different origins? Did different races come from different star systems? How many different humanoids are there? Legends have told us that we come from the Pleiades, Orion, Sirius, Cygnus, and elsewhere.

Could the earth itself have been terraformed for humans to exist here? The placement of earth in orbit, the tilt, spin speed, length of days, length of years. Were all these things engineered? What about the abundance of life?

The Moon

Scientists don't know where the moon came from or how it got to be in its orbit. There are two main theories of how the moon came to be. The first theory is capture. Capture means that the moon was slowly drifting by the earth and was captured into the earth's orbit. Most scientists don't agree with this theory because the physics is thought to be impossible. The second theory is collision. This means a large meteor hit the earth while the earth was forming, and the collision created the moon and then the moon stayed in earth's orbit. Although, the metallic composition of the rocks found on the moon are not from earth.

The moon's radius is 1079 miles, its distance from the earth on average is 238,900 miles, its orbital speed around the earth takes 27 days, it's average distance from the sun is 93 million miles. The moon's orbit period matches its rotation, so we only see one side of the moon.

The size of the moon and distance from the sun gives total eclipses. The sun's diameter is 400 times larger than the moon. The sun is also 400 times farther away from the moon. Scientists say that this placement could not happen by chance. The moon is 2160 miles around at its equator. It's 1/4th the size of earth. The moon helps to keep the tilt to the earth's axis steady, gives stability to the earth's wobble, gives the earth it's seasons, a gravitational pull to the earth's oceans, helps animals with their life cycles, and helps to keep eco-systems stable.

High tides are highest and low tides are lowest when it is a full moon or new moon because the gravitational pull of the moon and sun add up. When the moon is at perigee high tides are larger. At apogee the tides are lower.

The Zulu tribe from Africa have a legend that two gods made the moon and brought it into place. Wowane and Mpanku are the gods the Zulu shaman, Credo Mutwa says did this. Legend states that when the moon was brought into orbit, it caused floods, climate change, unusual weather and an axis shift. If the moon wasn't in its place the earth's wobble would become erratic and we wouldn't have seasons, poles would lose all their ice caps, and we would have extreme weather and climate changes. This all sounds like things that have happened to the earth in the ancient past. Was the moon put in its place? Who would have done this?

Mars

Mars is the fourth planet from the sun. Its distance from the sun is 141.6 million miles. One day on Mars is 24 hours and 37 minutes. The orbital speed around the sun is 53.7 thousand miles per hour and it takes 687 days to make one year. Mars is about half the diameter of Earth, it has 11% of Earth's mass, 15% of Earth's volume and 38% of Earth's gravity.

Scientists believe that Mars has a metallic core with a radius of about 1,794 km. The core is primarily iron, nickel and about 15% sulfur. The mantle is made of silicate. The crust also has silicon along with iron, magnesium, aluminum, calcium, and potassium. The crust is about 50 kilometers thick, and the planet is said to be seismically active.

In 1975 the Viking orbiter took pictures of Mars which gave geological evidence of dried upriver beds, lakes, and oceans. In 2004 NASA landed two rovers on Mars, Spirit and Opportunity. Each rover explored opposite sides of the planet. One rover took pictures of the landscape of Mars, and it was not red but looked like earth. The photos also showed a blue sky with an atmosphere. July 25th, 1976, Viking 1 took photos of the Cydonia region of Mars. The photos showed what is now called 'the face of Mars'. The anomaly looks like a large face 3 miles by 2 miles. The photo also showed what some believe to be pyramids near the face. A cartographer and satellite image expert Arel Torran mapped out the Cydonia region by mathematical relationships between the structures in the area. After studying the area, he concluded that the structures in the Cydonia region are set out to resemble the Pleiades star system. The tallest of these pyramidal structures is 2.5 miles high with a north south alignment with the axis of Mars. There was also said to be another structure nearby that resembles a sphinx. Many scientists and researchers think there might be a correlation with the pyramids and the sphinx on the Giza plateau in Egypt. In fact, the name Cairo, in Arabic, means 'place of Mars.'

Concentrated on one side of Mars is over 3000 craters from meteor hits. Some of them are 20 miles in diameter. Could these meteor hits have been caused by a planet which was destroyed in a planetary war?

When Viking went to Mars in 1976 Dr Gil Laven designed a test to detect life within the soil of the planet. The test was positive for microbes and life. NASA rejected the findings.

Mayan Destin Codex has a section written about Mars. Part of the codex brakes up days into 10 groups of 78 days to 780 days which is the time it takes Mars to get from one point on earth's horizon and back or synoptic of Mars. In Mayan tradition the Feathered Serpent was said to have killed Mars by stripping off its skin. Could the Mayan story be about a planetary war between Mars and a planet beyond Mars which is now the asteroid belt?

Approximately 65 million years ago Mars lost its atmosphere. Scientists found a layer of Zenon 129 which can only come from nuclear weaponry. Some scientists believe that Mars had oceans, rivers, lakes, and plant life at some point.

A day on Mars is 24 hours and 39 minutes which is a little longer than a day on earth. One year on Mars is 2 earth years because the orbit of Mars is twice as far from the sun as the orbit of earth. The tilt of Mars is 25 degrees and earth's tilt is 23.5 degrees. The tilt of Mars would also give seasons, support eco-systems, and oceanic tides just as the tilt of earth does. Could the tilt of Mars have been created to support a climate and eco system for human habitat in the past?

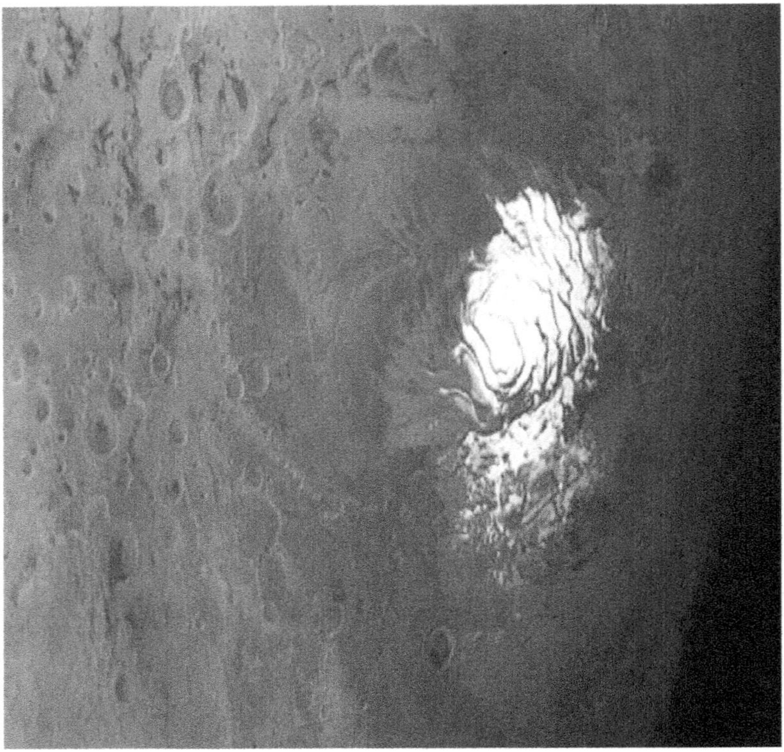

In March 2006, the Mars reconnaissance orbiter reached Mars. It was in orbit 186 miles above the surface and took 200,000 images. On sept 28th 2015 scientists say they found photo evidence of ice caps on the planet's poles. Also, geological evidence of water once being on Mars.

The Valles Marineris canyon on Mars is more than 4,000 km long, 200 km wide and up to 7 km deep. This area is close to the equator of the planet. Scientists say that it was not made by water or erosion. Some believe that the trench was made by an electrical discharge.

1974 in Pensse Journal Ralph E Jergens published an article called, 'Of the Moon and Mars'. In the article he stated that the trench looks like it was made by electrical discharge from advanced weaponry.

In Nov 2014 Ph. D John E Brandenburg wrote an article on proof of nuclear weapon signature on Mars. Zenon 129 was found, and that particular Zenon spectrum is only found from nuclear weaponry and not from nuclear energy production. The radiation patterns show 2 nuclear explosions sites that have global debris pattern. In the northern part of Mars Trinite or vitrified glass was also found.

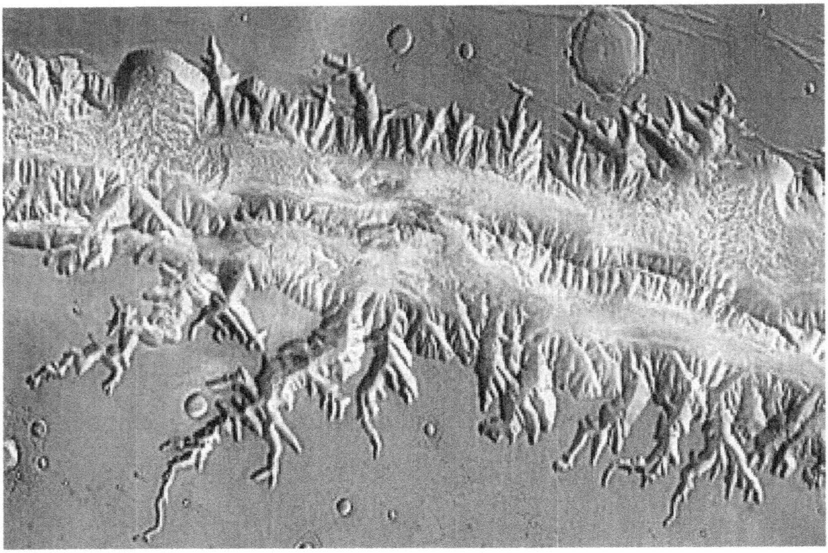

The Valles Marineris

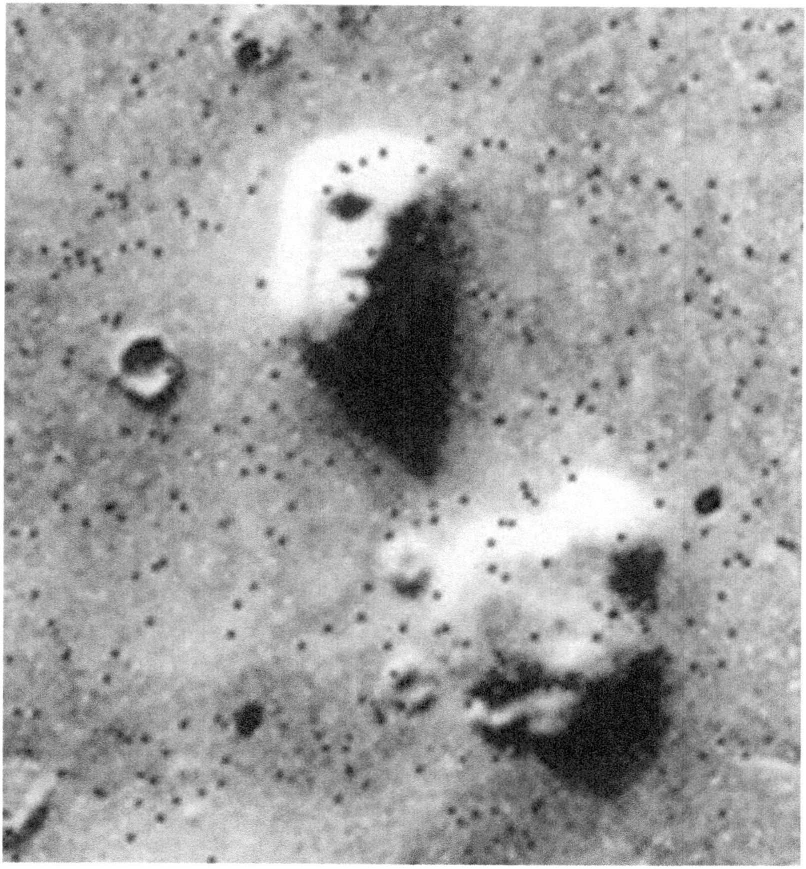

The Face of Mars

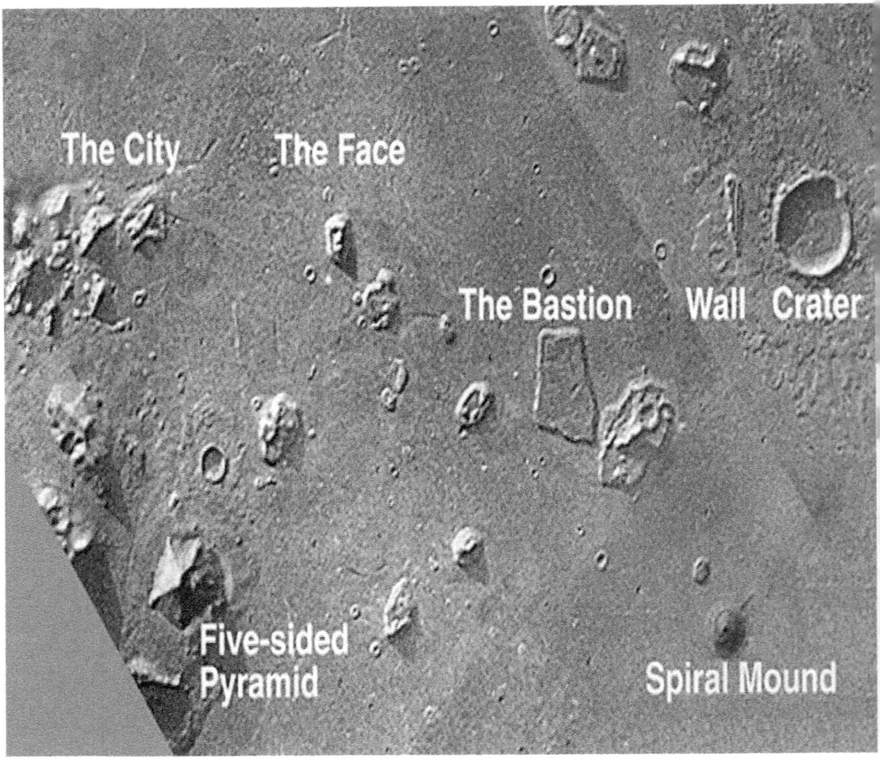

Cydonia region on Mars

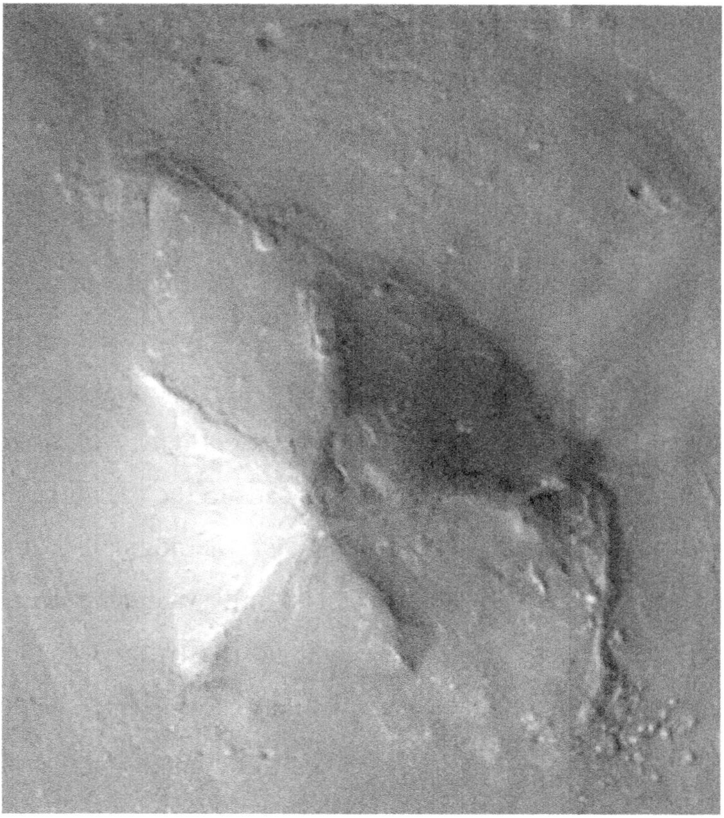

Five-sided pyramid on Mars

With a similar tilt, Mars may have had seasons much like Earth. The lengths of the seasons on Mars are about twice of those on Earth because the distance from Mars to the Sun is twice as far as the distance from the Earth to the Sun. The tempeture on Mars varies from highs up to 35 Celsius in summer at the equator and lows of -143 Celsius at the polar caps in winter. Mars receives approximately 43% of the suns light of Earth. Mars does fall into the habitual zone for human survival. Scientists have found evidence of flowing water on the surface of Mars. This leads to the idea that Mars could have contained life in the past. Geologically, at one time Mars had active Volcanos. The largest volcano is called Olympus Mons. It stands at 26 km and is 500 km at the base. Today scientists believe that Mars may have active volcanos. Vulcanism is necessary to recycle minerals into the environment and helps the eco-system to thrive. Nasa has found microbes on the surface of Mars, which implies that the planet could sustain simple life forms. The Phoenix lander found the Martian soil to be alkaline containing sodium, magnesium, potassium, and chlorine. These chemicals are all needed for life and the growth of plant life. A Mars orbiter also found large amounts of ice at both the poles of Mars. Scientists think that if the water in the polar caps where to melt it would be enough water to cover the surface of Mars 11 meters deep. In 2016 NASA also found ice in the Utopia region. It is believed to be equivalent to the amount of water in Lake Superior. The Korolev crater on Mars is said to hold ice which is 1.2 miles deep.

Mars lost its magnetosphere long ago. The solar winds now strip away atoms from the ionospheres outer layer. This creates a low surface pressure on Mars (0.6% of earth's) and low surface gravity (38% of earth's). scientist have found parts of the Martian crust to have been magnetized in a way that suggests a pole reversal in Mars's past and Tectonic activity.

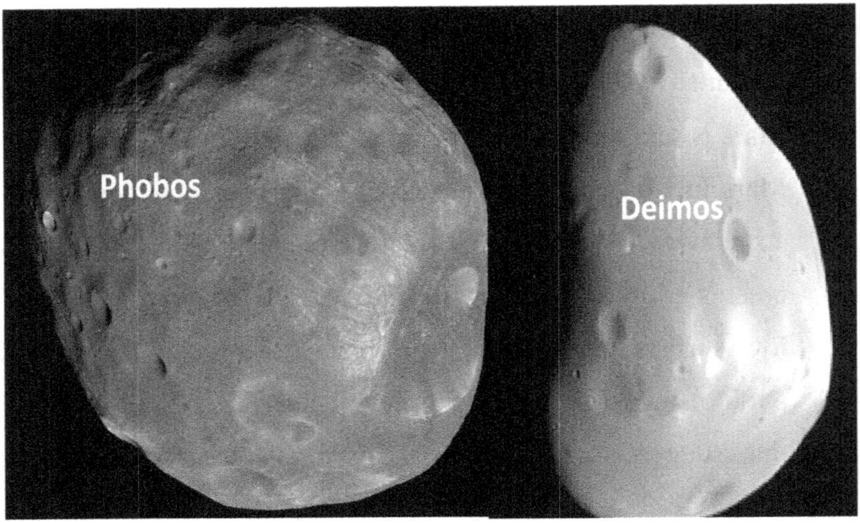

Phobos and Deimos, the moons of Mars

Mars has two moons, Phobos (about 22 km) and Deimos (about 12 km). Phobos rises in the west while Deimos rises in the east. The orbit of Phobos is deteriorating and will last approximately 50 million years before the moon crashes into Mars. The origin of the two moons is not known but they have a composition similar to asteroids. Their orbits are circular and near the equator which would be unlikely for captured objects given the requirements of the capture dynamics.

The days and seasons on Mars are like those on Earth. Both Mars and earth have similar tilt and rotational period. The atmosphere of Mars contains around 96% carbon dioxide, 2% argon, 2% nitrogen and traces of water and oxygen. Did Mars at one time have an atmosphere with life, water and eco-systems? The evidence gives way to the possibility. Could the moons of Mars have been put into place for the purpose of terraforming? If the structures photographed on Mars are not natural, then someone at some point in time made them. Was there a civilization that was destroyed by war on Mars?

The Asteroid Belt

The Asteroid belt is located between Mars and Jupiter. The total mass is estimated to be 2.39×10^{21} kilograms. Over half the mass of the belt is believed to be contained in four of the largest asteroids: Ceres, Pallas, Vesta and Hygiea. The asteroid belt is said to be made from a gas cloud which was not able to form a planet due to lack of gravity. Scientists of the day do not believe this theory.

In 1802 Dr. Olbers came up with the idea that Ceres and Pallas were pieces of a planet that occupied the region between Mars and Jupiter. Astronomer K.N. Savchenko introduced the possibility that Ceres, Pallas, Juno, and Vesta were all escaped moons. Savchenko thought that these four celestial bodies were moons of a planet that used to exist between the planets Mars and Jupiter. The Phaeton hypothesis was his idea that according to Bodies Law, there should be another planet called Phaeton were the asteroid belt is and that Ceres, Pallas, Juno, and Vests were moons that became homeless after the planet Phaeton was destroyed. In 1766 astronomer Johann Daniel Titus found a pattern in the orbits of the planets. He found a mathematical pattern now dubbed as Bodes Law. This pattern in the planets orbit that Titus found predicted a planet between Mars and Jupiter. Scientists now believe that the asteroid belt only contains a fraction of its original mass. The asteroid belt may in fact have had close to the same mass as the planet earth. It is thought that most of the material got ejected from the asteroid belt, leaving a very small percentage of its original mass. Could the asteroid belt have been a planet that was destroyed by war?

Sirius

Sirius is also known as the Dog Star. It is the brightest star in the Canis Major constellation. The Dogon tribe in Mali Africa say that a god named Namo visited their tribe and gave them knowledge. Namo told the Dogon that Amma, the sky god was Namo's creator. Namo is said to have come from Sirius. Namo told the Dogon tribe about the Sirius star system. He told the Dogon that there are three stars in the Sirius system: Sirius, Potollo (Sirius B) and Emme Ya (Sirius C). Namo told the Dogon that Sirius B was a collapsed star. He also shared the name of the planet in the star system where he came from, Xylanthia. The planet Xylanthia is said to have no night because of its three surrounding suns. The planet is believed to be the home of Osiris, the Egyptian god. It is also said to be where the people of Atlantis come from.

Visitors

In the Egyptian belief system, Zep Tepi is said to be a time when the sky gods came to earth. It is believed that Osiris came from Orion (the nursery of stars) and his consort Isis came from Sirius. The Dogon tribe of Africa believe that a god named Amma came from Sirius B. The Dogon called Sirius B, Potello. Sirius B wasn't known to modern science until the 1970's. The Zuni Indians from America say that sky beings taught them technology. The Zuni say Kachinas came from space through a portal. The Chinese tell a story about a space visitor called Wong Dee. Aboriginal people talk about visitors called Wongjinna, sky gods who created an eco-system here on earth. In New South Whales there is stone art of a visitor they call Oyammi, a creator god. Oyammi came from the morning star in the Milky Way. The story says, his space craft left a trail of fire. Mayans have visitor gods named 18 Rabbit and Pakal. Tikal (a city in Guatemala) is set to resemble the Pleiades. Mayans say they are from there. Mayans say a visitor came to them from the Pleiades called Kukulkan. He taught them technology and said he would return one day. The Hopi Native American Indians also say that a visitor came to them from the Pleiades. The Hopi believe that they themselves came from the Pleiades.

The Dogon say that a god named Namo came from the skies, descended, and gave them knowledge. Namo descended from the sky in an arc of fire. The Dogon have a ceremony in honor of this visit. Namo had an elongated head. In the 1920's French anthropologist Marcel Griaule went to Mali Africa to visit the Dogon. The Dogon told Griaule that Amma came from Sirius B which is undetectable without modern aid. Sirius is 8 light years away; it is the closest star to earth. Sirius B wasn't discovered unit the 1970's. This gives credit to the Dogon's visitation story. Sirius B is a white dwarf, partly burnt-out star, partly collapsed with a very dense mass, its weight is 8 times heavier our sun. Griaule discovered that the Dogon also had knowledge of the solar system of Sirius and the exact orbits of the stars in the system. Sirius A, B, C, and a planet named Xylanthia. Sirius A and B orbit Sirius C which is a black dwarf. Xylanthia is a planet that orbits Sirius C. Xylathia is said to have no night. Some stories claim that people from Atlantis came from Xylanthia.

The Native American Indian Zuni tribe from New Mexico left many cave drawings of what look like space men. The Zuni legend states that people descended from the sky. The elders called the visitors Kachinas. The Zuni say that the Kachina gods came down from the heavens to lead the Zunis to earth though a portal. The Zuni have a ceremony/celebration of when the Kachinas came to visit.

In China around 3000 B.C. it is said that Wong Dee came to the earth from the star Chi. The story says that Wong Dee was born from the belly of a fire breathing dragon. He was the first emperor in China. Legend also says that he came to earth on a flying dragon. He brought order and created an empire. He gave the people acupuncture, the compass, writing, and other knowledge. When he was done, he left on his dragon.

Mandeans from Persia are said to have a connection to Cygnus. The throne room of Abathur Muzania was made to resemble the Cygnus constellation. Nuit the Egyptian sky god is said to be from Cygnus. The Mayan sun god Kinich Ahau is said to have descended from Cygnus. The structures of Gobekli Tepi are aligned with Cygnus.

In Iraq, there is the Babylonian legend of Ahlish. In 1849 Sir Henry Leyard discovered the text about the Anunaki. The word Anunaki means; those whom from the heavens came. Some believe that the Anunaki came from Mars. Legend says that they had a supply base on Mars and a catastrophe happened that stripped away the atmosphere.

The Mayans built the city Tikal in Guatemala as a representation of the Pleiades star constellation. The Mayans believe that they came from the Pleiades. In the Mayan text, the Popol Vuh, it says that life was brought to earth by the gods from space. Bolon Yokte is a Mayan character who is said to have come from space to earth 900,000 years ago. This was written at Palenque in temple 14.

Mayans

Uaxaclajuun Ub'aah K'awiil

(Eighteen Rabbit)

Pakal

The Mayan civilization was a Mesoamerican civilization. They had a highly developed alphabet and writing system along with complex sciences like architecture, mathematics, astronomy, and an extremely accurate calendar system. They used Ch'olan and Yucatec languages. The Mayan writing system used hieroglyphs of about 500 characters with 200 of them being phonetic.

The Mayan civilization encompassed parts of El Salvador, Honduras, Guatemala and Belize from about 2000 B.C. to 250 A.D. The religion of the Mayan stated that a divine King would mediate between mortals and gods.

Building materials where usually volcanic tuff, sandstone, fired bricks, and obsidian. There were many pyramids in the Mayan territory. The pyramids' purpose and function are unknown but many of them are aligned with celestial movements. The Caracol at Chichen Itza seemed to track the motions of the planet Venus. Scientists found structures that they call E Groups. E Groups are temples that consist of three small structures which face a fourth structure. The E Group structures track the solstices and equinoxes.

The Mayan numbering was a base 20 numbering system. It was expressed using a bar and dot writing scheme. They also used the value of zero before any other known civilization. Stacking these symbols upwards when writing them meant that the value of the symbol went up. The symbol on the bottom has a value 0 to 9. The symbol above that is a multiple of twenty. The symbol above that is a multiple of 400 and so on. This made large calculations easy by just adding bars and dots in similar columns.

The Mayan had the most complex calendar system in Mesoamerica. It is known to be more accurate than the Julian calendar. They had developed a way of telling time using cycles of the sun, earth and moon. Using a analog computer of sorts, they constructed three interlocked niched gears that turned together. One gear has a 260 count, one has a 360 count, and one has a 52 count. For an unknown reason the Mayan started their calendar in 3114 B.C. The Mayans recorded the movements of many planets in their codex including Mercury, Mars, Jupiter, and Venus.

Hopi

A Kachina doll

The Hopi Indians in North America believe that they came from the Pleiades star system. The Hopi Native American Indians live in Northeastern Arizona, and at one time they inhabited the four conners region of the United States. The four conners includes parts of Utah, Colorado, New Mexico, and Arizona. The Hopi didn't believe in land ownership or dividing the land. They also believe that time was happening in the past, present, and future all at once. Their religion is a moral and ethical spirituality. The word Hopi is said to mean behaving one, peaceful ones, polite ones. Hopi ceremonies follow the cycles of the moon. Hopi have gods who came to them from space and gave them knowledge and religion. These beings are called Kachinas.

Zuni

The Zuni are an American Indian tribe from New Mexico. Historically they lived along the Zuni River in the Zuni Pueblo 34 miles south of Gallup New Mexico. The Zuni lived with and where friendly with the Anasazi at Chaco Canyon around 1100 A.C. Later they made villages on top of mesas and lived along the Rio Grande. There is no known relationship with the Zuni language to other Native American language. However, a few words having to do with religion are shared between the Zuni and the Keresan, and the Pima and Hopi Native American Indians. Religion to the Zuni involved belief in three most powerful deities: Earth, Sun, and Moon. Their ceremonies followed the cycles of the sun, for example the Winter solstice ceremony.

Dogon

The Dogon are a tribe in the central plateau region of Mali in West Africa. The Dogon language is not related to any other language in the region. The people have isolated themselves from other villages and do not except the prevailing Islamic religion.

French anthropologist Marcel Griaule lived with the Dogon for fifteen years, then in 1946 he met with Dogon elder Ogotemmeli. The elder told Griaule about Dogon symbols and religion. It was found that the Dogon had knowledge of astronomy, measurements of calendrical time, and pharmaceutical medicines. The Dogon religion is very precise. Their god Nummo and twin Nommo came from the Sirius star system. Ogotemmeli described the Nummo as serpent like with green skin and scales like a fish or snake that could reflect different colors. The Nummo were said to walk upright and to have tails. The Nummo gods were hermaphrodites. Nummo told Ogotemmeli that the tradition of twin and hermaphrodite birth of humans had come to a stop. At some point in the past, humans began being born as single sex and without a twin. When twins are born in the Dogon community, it is celebrated as a tribute to the past when all humans were born into life as a pair.

During Griaule's time with Ogotemmeli, he found that the Dogon had knowledge of were their ancestors came from. Griaule learned about Sirius (Sigi Tolo), Sirius B (Po Tolo), and Sirius C (Emme Ya Tolo). Sirius B takes 50 years to orbit Sirius and Dogon ceremonies celebrate this event. The Dogon believe that Sirius C is a red or brown dwarf star. This is still to be proven by modern day science. The ceremonies of the Dogon include dancers who wear masks that resemble the Nummo. Walu and Satimbe masks seem to resemble the Kachina masks and dancers of the Hopi Native American Indian gods.

Osiris

Osiris is the Egyptian god of fertility, God of the dead, and resurrection. Legend says that after his death, he was set among the stars as the constellation Orion. The story of Osiris comes from the pyramid texts and the Shabaka Stone. The name is from Greek and Latin. Egyptian hieroglyphic writing gives the name as 'Wsjr'. The myth describes Osiris being killed by his brother the Egyptian god Set. His wife Isis gathers his body parts, then conceives and gives birth to their son Horus. Osiris is usually depicted wearing his Atef Crown which symbolizes the sleeping mind or the unconscious mind.

Egypt, Atlantis and Xylanthia

The Greek philosopher Solon (630 – 560 B.C.) went to Egypt to learn about ancient wisdom from the Egyptian priests. The priest Sais told Solon that approximately 13,000 years ago a city called Atlantis was destroyed. The priests left there and came to Egypt to start a new civilization. Solon was also told that the people from Atlantis came from a place called Xylanthia which was in the Sirius star system. Xylanthia was said to orbit the star Sirius C which is a black dwarf. Did the people from Xylanthia come to earth when Sirius C burned out and became a black dwarf? Plato later wrote about Atlantis and said that the king of Atlantis was called Atlas. In Mauritania Africa there is a place known as the Richat structure. It is also known as the 'Eye of Africa'. The local legend states that the name of the first King of Mauritania was a man called Atlas. Could this be the same person Plato spoke of? Herodotus mapped the known world in 450 B.C. His map included the Mauritania region of Africa where the Richat structure is now. There in this region on his map it says Atlantis. The Richat structure is made of 3 concentric circles or rings, just as Plato described. The Richat structure is 23.5 km across, it also has an outlet leading to the Atlantic Ocean. These details match with Plato's descriptions of Atlantis.

The Richat Structure

Cygnus

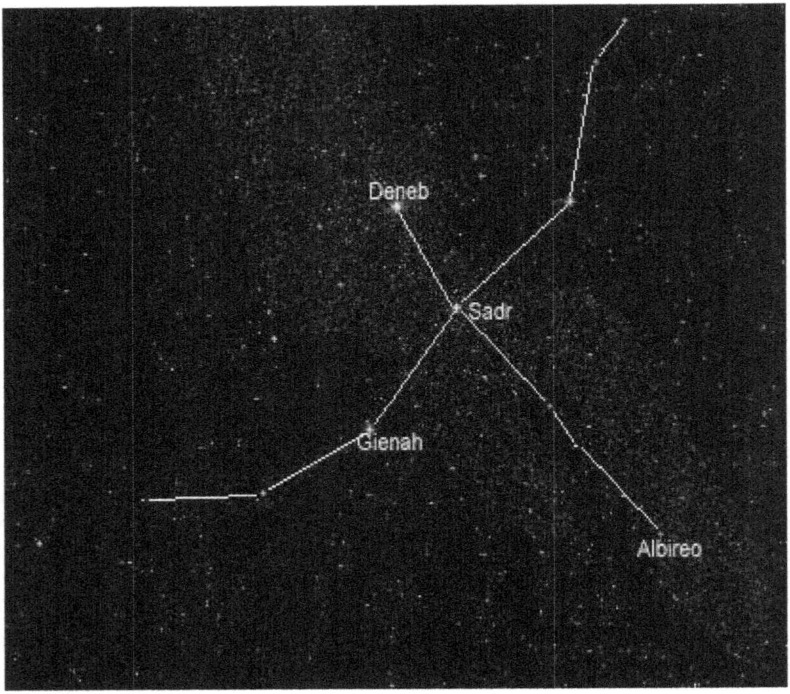

Cygnus is a constellation in the northern hemisphere. The name Cygnus means swan. The constellation is made of 6 stars: Albireo, Aljanah, Azelfafage, Deneb, Fawaris , and sadr. There are 10 stars within Cygnus that have known planets orbiting them. Cygnus is in the fourth quadrant of the northern hemisphere. It is visible between latitudes +90° and -40°. The constellations neighboring Cygnus are Cepheus, Draco, Lacerta, Lyra, Pegasus, and Vulpecula. In Latin, Cygnus means 'the swan', and in Greek mythology Zeus set King Kolonai among the stars as the swan constellation Cygnus. The constellation is also known as the Northern Cross.

Deneb is the brightest star in Cygnus. It is a white super giant and is approximately 3200 light years away. Albireo is an orange giant and is binarily paired with a blue green hued star and is 380 light years away. Sadr is a yellow super giant. It is about 1500 light years away.

Fawaris is a blue giant and is about 171 light years away. Aljanah is an orange giant and is 72 light years away.

There are several X-ray sources in Cygnus such as Cygnus X-1 which is the first x-ray source thought to be a black hole. Cygnus is known as the Swan constellation or the Northern Cross. Fawaris and Aljanah are the wings and Deneb is the tail and Albiree is the beak or tip of the cross. Deneb, Albireo and Sadr forming the pole of the cross and Fawaris and Aljanah forming the arms of the cross.

During its extrasolar search for planets, the Kepler satellite survey concluded that there are approximately 100 stars with plants orbiting them. The number of stars that are found with planets orbiting them is continuously climbing.

Pleiades

The Pleiades constellation, also known as the Seven Sisters from mythology, is in the Northern Hemisphere. There are nine bright stars in the Pleiades constellation: Sterope, Merope, Electra, Taygeta, Maia, Celaeno, Alcyone, along with Atlas and Pleione. Its name means the sailing ones.

Sirius

Sirius is the brightest star in the night sky. Also known as Alpha Canis Majoris (Latin), Giving its nickname 'The dog star'. It's Greek meaning, 'glowing' or 'scorching'. The star Sirius is binarily paired with another star termed Sirius B. Sirius B became a red giant then collapsed into a white dwarf about 120 million years ago. The two orbit each other every 50 years. It is now believed that there could be a Sirius C which also orbits Sirius. Sirius is moving closer to earth, and some say that our sun and Sirius orbit each other every 26,000 years. It's heliacal rising in our night sky is approximately 365.25 days.

A Solar Planetary War

There is evidence of ancient wars all over the world and on Mars. On earth there are many ruins that show this evidence. What is seemingly erosion, is now thought to be signs of destruction by means of advanced weaponry. The ruins are still here today for us to examine. The stories of the ancients are here for us to examine as well.

Pachacute, an Incan king was believed to hold magical powers, or possibly advanced technology. He was known as a warrior king, in one of his battles it was said that he made boulders float and caused them to rain down on his enemies killing them. Atilla the Hun was said to be able to control the weather. Legend states that he could cause it to snow and cause extreme weather to help defeat enemies in battle. There are several stories about Vikings having advanced technology. The Viking Freyr was believed to have control over the weather. Gungnir, Odin's spear, is said to be a high-tech weapon that always hit its target like a guided weapon. Thor's hammer, Mjolnir, was said to bring down lightning and was able to destroy an entire mountain. In Celtic legend there was a god named Lugh also known as the Shinning One and the god of war. He was said to have had a magic spear of lightning, and it is said to have a roar like thunder.

Ancient Hindu scriptures speak of wars fought with high tech weapons such as lasers, missiles, nuclear bombs, etc. Vishnu had missiles which were motion detecting and heat seeking. In the Mahabharata 46 different types of weapons are described such as stealth technology, chemical warfare, and a weapon called Brahmastra which could destroy planets. In Cambodia, the god Preah Pisnokar was said to have a sword of light that could cut through anything much like a light saber. He is believed to have created the city of Ankor Wat .

The Mahabharata was written 12,000 years ago (some say it's much older), it speaks of the god Rama. Rama tells a story about ancient wars fought with weapons of advanced technology. One weapon called Brama was said to be similar to a nuclear explosion and caused what was described to be radiation sickness. The explosion of the Brama was said to be brighter than a thousand suns. Trees were said to burn, and people lost their hair and fingernails afterwards. In 1920 there was evidence found of a nuclear explosion in the Pakistan Indus valley when the ruins of Mohenjo Daro were discovered. High levels of radiation were found and areas that had been vitrified.

The Hindu text, Bavahagita, states in detail a weapon as a bolt of iron charged with the light of a thousand suns. The Bavahagita also speaks of radiation sickness, animals dying, and the shadows of people who died from the blast. The Bavahagita also describes flying crafts called Vimanas.

Karnak Egypt

Karnak temple in Egypt is another temple left in ruins. How did this megalithic structure (like many others) and its massive stones become cracked, burnt, broken, and thrown about? Scientists are coming to the conclusion that many of the worlds megalithic structures have been purposely and methodically destroyed. Some structures look as if they have been exploded and many have even been transmuted or experienced vitrification. How could these massive stone structures, most made of the hardest kind of stones, been destroyed in this way? Could it have been an energy weapon of advanced design?

Karnak Egypt

Karnak Egypt

Above we see hard granite stone dissolving away. Below we see granite which has been cracked.

Karnak Egypt

Dendera Egypt

Above we see more cracked granite. This alter was made from one piece of solid granite stone.

Puma Punku

Above is the gateway found at Puma Punku. The crack in the gateway and the way in which the massive stone blocks in the picture below have been thrown about does not seem to have happened from a natural form of erosion.

Puma Punku

Cuzco Peru

Above is a photo of the famous stairwell at Cuzco Peru. It is a massive megalithic stone with a stairway cut into it which has been thrown upside down. It has been burnt, cracked and it looks like it has turned to glass in places. How could this have happened? It would take an enormous amount of energy to lift this stone and set it upside down.

Scotland

This structure in Scotland is famous for the fact that the stones have gone through the vitrification process. The ancient site shows evidence of a nuclear blast.

Scotland

Mexico

Above is the Pyramid of the feathered serpent at Teotihuacan in Mexico. As shown, the top of the pyramid has been destroyed and rebuilt in later times. Below is a structure that has been burnt and destroyed.

Mexico

Mexico

Above is the Pyramid of The Moon. Here you can see where the stairway ends, the top of the pyramid has been rebuilt. The top of the pyramid was destroyed. Were these pyramids in Mexico hit from above from an energy weapon?

Stonehenge England

Above is Stonehenge. We can see the tallest stone in the picture has a point sticking out on top. These points lock the top stones into place. The stone that lay on top are grooved to set onto the points so that the top stones cannot simply be pushed off or slide about. This site is very peculiar in that so many of the stones in the henge are missing. Like many ancient ruins, Stonehenge has been restored to some degree. Older pictures show Stonehenge with some of these large stones thrown about or pushed over by some unknown force.

Karnak

Above is the Temple of Karnak. Below the Temple of Luxor. These temples are but a few miles apart. Both these temples have had their roof destroyed and have stones which have been thrown about. Did these temples get hit from above by an energy weapon?

Luxor

Giza

Above is the Sphinx Temple and below is the Osirian. Both temples have had their roofs destroyed. Again, many of the megalithic stones have been thrown about.

Osirian

China

Above is one of many pyramids in China. It has been covered by years of dirt and soil. Many pyramids (and other ancient sites) around the world have been buried in soil and earth. Some scientists think these sites were buried by an ancient flood. Some scientists think the site were buried on purpose. Could these sites have been buried by and explosion cause by a weapon? Could a massive explosion have thrown up a large amount of earth and dirt into the atmosphere? Could all this dirt could have then come back down and buried these sites afterwards?

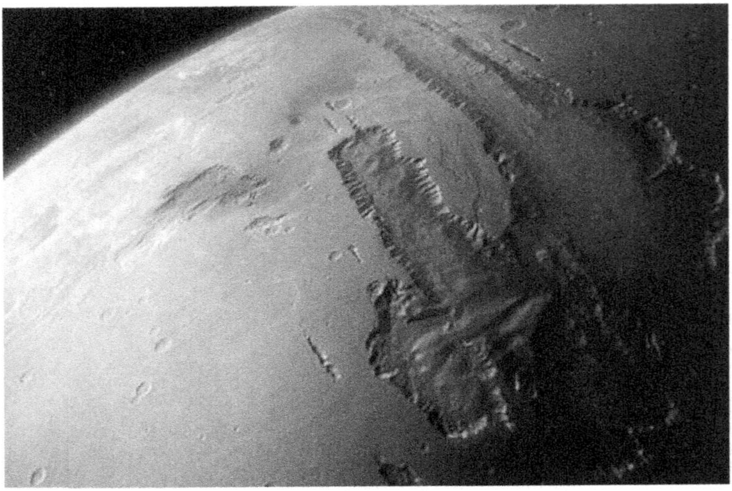

Mars

The Valles Marineris on Mars is believed to have been caused by some sort of energy weapon.

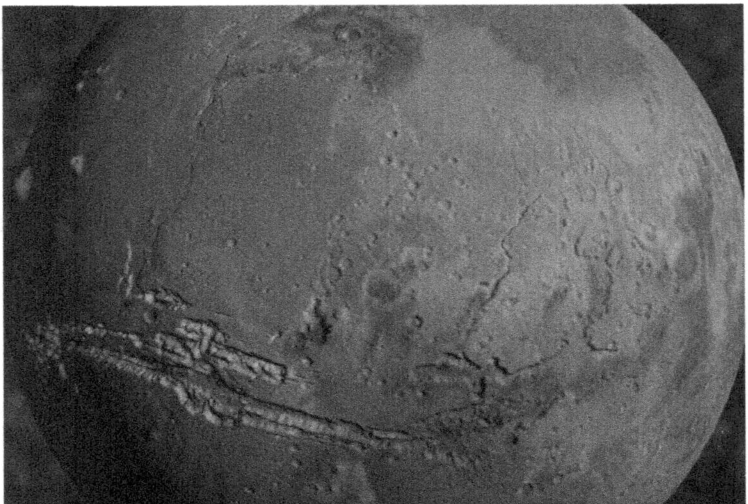

The planet Mars was found to have a large amount of Zenon 129 which is the unmistakable signature of nuclear weaponry. This particular element can only be produced from a nuclear reaction that happens from a nuclear weapon.
The Valles Marineras on Mars is 4000 km long and up to 7 km deep. Scientists say it was not made by running water. This canyon looks like it was made by an electrical discharge. Could this also be evidence of weaponry? Possibly an energy weapon? If Mars were destroyed by weapons and there is evidence that a planet beyond Mars exploded, could there have been a planetary war between the two?

The Mayan kept careful records on the planet Mars. There is a section about Mars in the Mayan Destin Codex which tracks the synoptic period of Mars. The codex breaks up the days of Mars into 10 groups of 78 days. Approximately 65 million years ago, Mars lots its atmosphere. This also coincides with the extinction of the dinosaurs.
In Mayan tradition the Feathered Serpent was said to have killed the god Mars by stripping off its skin. Does this Mayan legend tell the story of Mars losing its atmosphere in a war?

In 1974 Pensse Journal published an article by Ralph E Jergens called 'Of the Moon and Mars'. Jergens said that the Valles Marineras looked like an energy weapon discharge. In 2014 Ph.D. John E Brandenburg wrote an article that provided evidence of a nuclear weapon use on the planet Mars. Brandenburg spoke of the Zenon 129 signature found which can only be produced from a nuclear weapon. The radiation shows two nuclear explosion sites that have global debris pattern. Trinitite or vitrified glass was also found in these regions.

The Asteroid Belt

Could the Asteroid Belt have been a planet that was destroyed in a planetary war? Many people believe that a civilization on Mars and the planet beyond it started a war in which both planets were lost. The orbit of the asteroid belt is precisely where a planet should be. In 1766 Johann Daniel Titus, and in 1772 Johann Elert Bode found a mathematical correlation in the orbits of the planets around the sun. Now known as Bodes Law, the planets orbits are spaced out in a precise distance to one another. According to Bodes Law there should be a planet in orbit where the asteroid belt is. There is evidence of a planet's destruction as well. The moon of earth is tidally locked into orbit. One side of our moon always faces the earth, and one side of our moon always faces out towards space. The side facing out has a massive amount of craters from asteroid hits. Could all or most of these collisions happened at once? The moon of Saturn Lepidus also has one side that always faces outward into space. Lepidus has burn marks on the side that faces out as if a massive explosion spit fire and gasses into its surface blackening Lepidus. Could this have happened from a planet that exploded and is now the asteroid belt? On earth, the Thiexulub crater in the Yucatan peninsula was 6 miles in diameter, the Boltish crater in Ukraine is 15 miles in diameter, the Silver Pit crater in the North Sea is 12 miles in diameter, and the Shiva crater in India is 300 miles in diameter, all happened around the same time period.

The K-T boundary recorded this period in the earth's soil. Is it possible that these asteroids that hit earth at this time where from a planet beyond Mars that exploded?

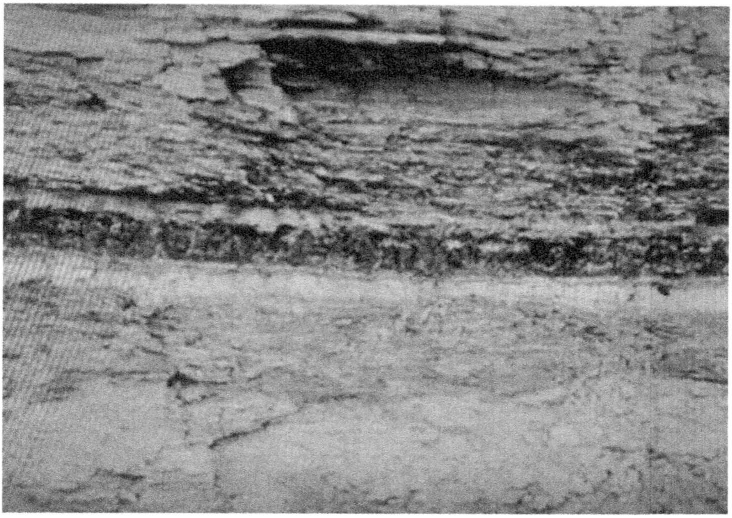

The K-T Layer

Could the K-T Layer and other meteor hits be evidence of this planet's destruction?

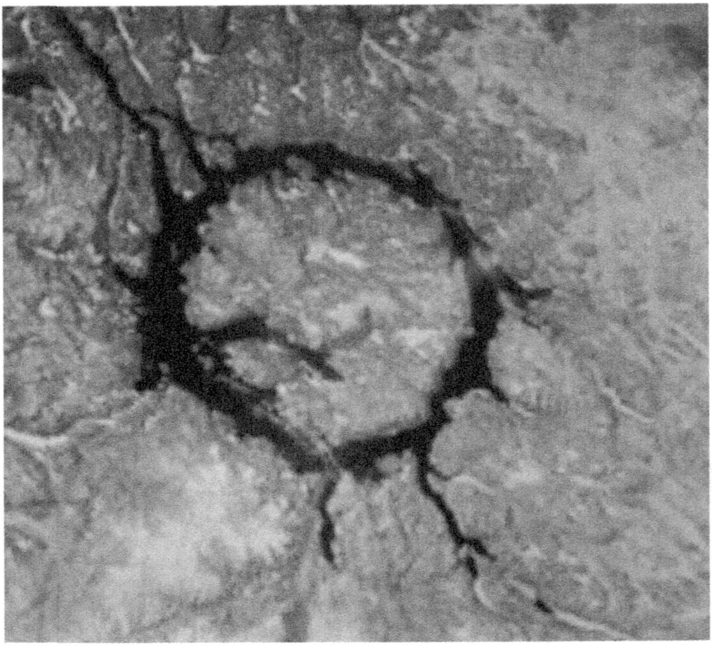

Shiva Crater

Boltish Crater Ukraine

All these craters (and more) are thought to have been made around the same time.

Silver Pit Crater North Sea

Crater Hits to our Moon

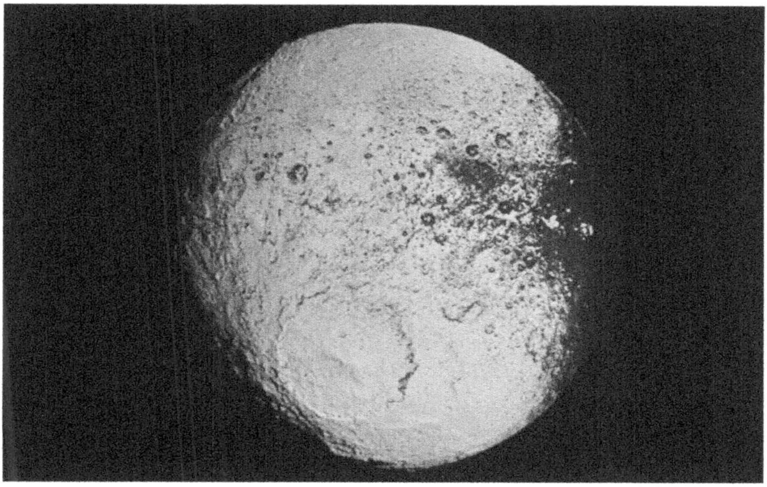

Saturn's Moon Lupidus

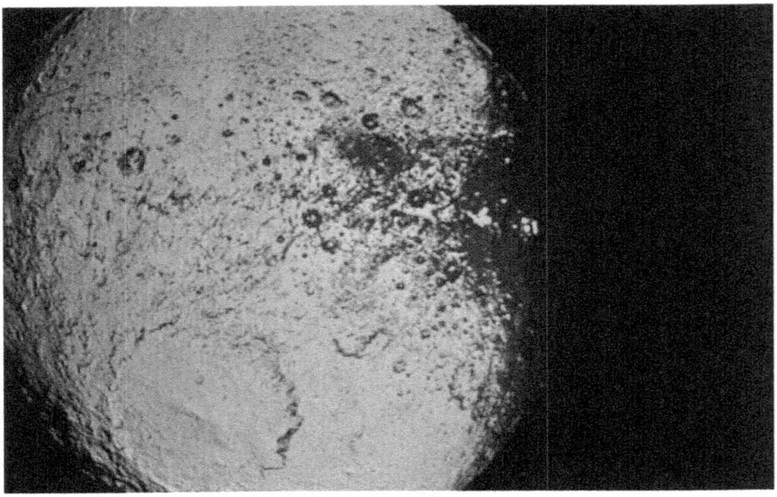

Lupidus

Both moons show meteor hits on the one side that always faces outward away from their planets. Could these meteor hits be from the planet just beyond Mars which was destroyed in a planetary war? Mars also has a large number of meteor hits on one side of its surface.

Summary

There is a great amount of evidence that life did not originate on the planet earth or any other planet in this solar system. There is also evidence that there was highly advanced ancient technology on this planet and possibly others such as Mars. Starting with the solar system itself. Could the planets have been placed around our sun in their order? Could their moons have been put into place to give the planets their tilts, seasons, and to help eco-systems thrive? Earth does seem to have been terraformed for life. The amount of water is hard to account for as well as the abundance of animal and plant life. The fossil records show that animals appear and do not show signs of evolution, as if they were brought here. So many different types of eco-systems in a perfect delicate balance supporting an array of life. The weather, climate, oxygen level, atmospheric pressure at different levels, and an infinite number of other variables at optimal conditions for so much life. There have been several Homo-species living at the same time along with Homo-sapiens and no fossil record of any species giving rise to Homo-sapiens.

Mars has evidence of past life. Could there have been living entities there long ago, and perhaps the planet beyond Mars which is now the asteroid belt? There may have been a war that destroyed both planets. If this did happen, where are those warring parties? Did they kill each other off? Did they leave the solar system? Are they coming back some day? Are they, our ancestors? Most all the ancient megalithic sites on earth that show possibilities of advanced technology have been destroyed a very long time ago. Who was responsible for this destruction, and why was it done? Will the ones responsible for this destruction be a threat to us in the future? There have been numerus stories from every race on the planet earth that we have been visited by beings that have advanced technology. These visitors are said to be from different constellations and solar systems. Some of these stories say that these space visitors brought life to this planet and even placed our moon in orbit. Could all this be true? These stories even include a war that took place on Mars. If all these things are true, then the conclusion could only be that humanity did not come from earth. The next question is, where did we come from and how did we get here? Are we the aliens?

Made in the USA
Monee, IL
04 November 2026

67380062R00066